SHIT

TO REMEMBER

INTERNET ADDRESS AND PASSWORD KEEPER TO PREVENT WTF MOMENTS

D1530252

sourcebooks

Copyright © 2022 by Sourcebooks
Cover and internal design © 2022 by Sourcebooks
Cover images © macrovector/Freepik
Internal images © Makyzz/Freepik

Sourcebooks and the colophon are registered trademarks of Sourcebooks.

Published by Sourcebooks
P.O. Box 4410, Naperville, Illinois 60567-4410
(630) 961-3900
sourcebooks.com

Printed and bound in the United States of America.
VP 10 9 8 7 6 5 4 3 2 1

THIS BOOK BELONGS TO

NETWORK SETUP

All the shit you need to keep your internet up and running

HOME NETWORK SETTINGS

BROADBAND MODEM	
MODEL	
SERIAL NUMBER	
MAC ADDRESS	
ADMINISTRATION URL/IP ADDRESS	
WAN ADDRESS	
USERNAME	
PASSWORD	

WI-FI SETTINGS	
NETWORK NAME (SSID)	
CHANNEL	
SECURITY MODE	
SHARED KEY (WPA)	
PASSPHRASE (WEP)	

ROUTER/WIRELESS ACCESS POINT

MODEL	
SERIAL NUMBER	
DEFAULT IP ADDRESS	
DEFAULT USERNAME	
DEFAULT PASSWORD	
USER DEFINED (CUSTOM) IP ADDRESS	
USER DEFINED (CUSTOM) USERNAME	
USER DEFINED (CUSTOM) PASSWORD	

WAN SETTINGS

MAC ADDRESS	
IP ADDRESS	
HOST NAME	
DOMAIN NAME	
SUBNET MASK	
DEFAULT GATEWAY	
DNS—PRIMARY	
DNS—SECONDARY	

LAN SETTINGS

IP ADDRESS	
SUBNET MASK	
DHCP RANGE	

COMPUTER AND EMAIL SETUP

Insert a damn bookmark, these are the big ones!

COMPUTER MODEL/BASIC INFORMATION	
MODEL NUMBER	
ACCOUNT NUMBER	
TECH SUPPORT (PHONE #)	
CUSTOMER SERVICE (PHONE #)	

PERSONAL EMAIL	
WEBSITE/APP NAME	
USERNAME	
PASSWORD	

WORK EMAIL	
WEBSITE/APP NAME	
USERNAME	
PASSWORD	

OTHER/ALTERNATIVE EMAIL

WEBSITE/APP NAME	
USERNAME	
PASSWORD	

PASSWORDS

*Sorted alphabetically to make things
as easy as fucking possible*

A	
NAME/WEBSITE	
USERNAME/LOGIN	
PASSWORD	
NOTES	
NAME/WEBSITE	
USERNAME/LOGIN	
PASSWORD	
NOTES	
NAME/WEBSITE	
USERNAME/LOGIN	
PASSWORD	
NOTES	

A
NAME/WEBSITE
USERNAME/LOGIN
PASSWORD
NOTES
NAME/WEBSITE
USERNAME/LOGIN
PASSWORD
NOTES
NAME/WEBSITE
USERNAME/LOGIN
PASSWORD
NOTES
NAME/WEBSITE
USERNAME/LOGIN
PASSWORD
NOTES
NAME/WEBSITE
USERNAME/LOGIN
PASSWORD
NOTES
NAME/WEBSITE
USERNAME/LOGIN
PASSWORD
NOTES

	A
NAME/WEBSITE	
USERNAME/LOGIN	
PASSWORD	
NOTES	
NAME/WEBSITE	
USERNAME/LOGIN	
PASSWORD	
NOTES	
NAME/WEBSITE	
USERNAME/LOGIN	
PASSWORD	
NOTES	
NAME/WEBSITE	
USERNAME/LOGIN	
PASSWORD	
NOTES	
NAME/WEBSITE	
USERNAME/LOGIN	
PASSWORD	
NOTES	
NAME/WEBSITE	
USERNAME/LOGIN	
PASSWORD	
NOTES	

B

NAME/WEBSITE

USERNAME/LOGIN

PASSWORD

NOTES

NAME/WEBSITE

USERNAME/LOGIN

PASSWORD

NOTES

NAME/WEBSITE

USERNAME/LOGIN

PASSWORD

NOTES

NAME/WEBSITE

USERNAME/LOGIN

PASSWORD

NOTES

NAME/WEBSITE

USERNAME/LOGIN

PASSWORD

NOTES

NAME/WEBSITE

USERNAME/LOGIN

PASSWORD

NOTES

NAME/WEBSITE

USERNAME/LOGIN

PASSWORD

NOTES

NAME/WEBSITE

USERNAME/LOGIN

PASSWORD

NOTES

NAME/WEBSITE

USERNAME/LOGIN

PASSWORD

NOTES

NAME/WEBSITE

USERNAME/LOGIN

PASSWORD

NOTES

NAME/WEBSITE

USERNAME/LOGIN

PASSWORD

NOTES

NAME/WEBSITE

USERNAME/LOGIN

PASSWORD

NOTES

B
NAME/WEBSITE
USERNAME/LOGIN
PASSWORD
NOTES
NAME/WEBSITE
USERNAME/LOGIN
PASSWORD
NOTES
NAME/WEBSITE
USERNAME/LOGIN
PASSWORD
NOTES
NAME/WEBSITE
USERNAME/LOGIN
PASSWORD
NOTES
NAME/WEBSITE
USERNAME/LOGIN
PASSWORD
NOTES
NAME/WEBSITE
USERNAME/LOGIN
PASSWORD
NOTES

NAME/WEBSITE	
USERNAME/LOGIN	
PASSWORD	
NOTES	

NAME/WEBSITE	
USERNAME/LOGIN	
PASSWORD	
NOTES	

NAME/WEBSITE	
USERNAME/LOGIN	
PASSWORD	
NOTES	

NAME/WEBSITE	
USERNAME/LOGIN	
PASSWORD	
NOTES	

NAME/WEBSITE	
USERNAME/LOGIN	
PASSWORD	
NOTES	

NAME/WEBSITE	
USERNAME/LOGIN	
PASSWORD	
NOTES	

C

NAME/WEBSITE	
USERNAME/LOGIN	
PASSWORD	
NOTES	

NAME/WEBSITE	
USERNAME/LOGIN	
PASSWORD	
NOTES	

NAME/WEBSITE	
USERNAME/LOGIN	
PASSWORD	
NOTES	

NAME/WEBSITE	
USERNAME/LOGIN	
PASSWORD	
NOTES	

NAME/WEBSITE	
USERNAME/LOGIN	
PASSWORD	
NOTES	

NAME/WEBSITE	
USERNAME/LOGIN	
PASSWORD	
NOTES	

NAME/WEBSITE

USERNAME/LOGIN

PASSWORD

NOTES

NAME/WEBSITE

USERNAME/LOGIN

PASSWORD

NOTES

NAME/WEBSITE

USERNAME/LOGIN

PASSWORD

NOTES

NAME/WEBSITE

USERNAME/LOGIN

PASSWORD

NOTES

NAME/WEBSITE

USERNAME/LOGIN

PASSWORD

NOTES

NAME/WEBSITE

USERNAME/LOGIN

PASSWORD

NOTES

C

NAME/WEBSITE	
USERNAME/LOGIN	
PASSWORD	
NOTES	

NAME/WEBSITE	
USERNAME/LOGIN	
PASSWORD	
NOTES	

NAME/WEBSITE	
USERNAME/LOGIN	
PASSWORD	
NOTES	

NAME/WEBSITE	
USERNAME/LOGIN	
PASSWORD	
NOTES	

NAME/WEBSITE	
USERNAME/LOGIN	
PASSWORD	
NOTES	

NAME/WEBSITE	
USERNAME/LOGIN	
PASSWORD	
NOTES	

C

C

NAME/WEBSITE

USERNAME/LOGIN

PASSWORD

NOTES

NAME/WEBSITE

USERNAME/LOGIN

PASSWORD

NOTES

NAME/WEBSITE

USERNAME/LOGIN

PASSWORD

NOTES

NAME/WEBSITE

USERNAME/LOGIN

PASSWORD

NOTES

NAME/WEBSITE

USERNAME/LOGIN

PASSWORD

NOTES

NAME/WEBSITE

USERNAME/LOGIN

PASSWORD

NOTES

D

NAME/WEBSITE	
USERNAME/LOGIN	
PASSWORD	
NOTES	
NAME/WEBSITE	
USERNAME/LOGIN	
PASSWORD	
NOTES	
NAME/WEBSITE	
USERNAME/LOGIN	
PASSWORD	
NOTES	
NAME/WEBSITE	
USERNAME/LOGIN	
PASSWORD	
NOTES	
NAME/WEBSITE	
USERNAME/LOGIN	
PASSWORD	
NOTES	
NAME/WEBSITE	
USERNAME/LOGIN	
PASSWORD	
NOTES	

NAME/WEBSITE

USERNAME/LOGIN

PASSWORD

NOTES

NAME/WEBSITE

USERNAME/LOGIN

PASSWORD

NOTES

NAME/WEBSITE

USERNAME/LOGIN

PASSWORD

NOTES

NAME/WEBSITE

USERNAME/LOGIN

PASSWORD

NOTES

NAME/WEBSITE

USERNAME/LOGIN

PASSWORD

NOTES

NAME/WEBSITE

USERNAME/LOGIN

PASSWORD

NOTES

D

NAME/WEBSITE	
USERNAME/LOGIN	
PASSWORD	
NOTES	

NAME/WEBSITE	
USERNAME/LOGIN	
PASSWORD	
NOTES	

NAME/WEBSITE	
USERNAME/LOGIN	
PASSWORD	
NOTES	

NAME/WEBSITE	
USERNAME/LOGIN	
PASSWORD	
NOTES	

NAME/WEBSITE	
USERNAME/LOGIN	
PASSWORD	
NOTES	

NAME/WEBSITE	
USERNAME/LOGIN	
PASSWORD	
NOTES	

D

NAME/WEBSITE	
USERNAME/LOGIN	
PASSWORD	
NOTES	

NAME/WEBSITE	
USERNAME/LOGIN	
PASSWORD	
NOTES	

NAME/WEBSITE	
USERNAME/LOGIN	
PASSWORD	
NOTES	

NAME/WEBSITE	
USERNAME/LOGIN	
PASSWORD	
NOTES	

NAME/WEBSITE	
USERNAME/LOGIN	
PASSWORD	
NOTES	

NAME/WEBSITE	
USERNAME/LOGIN	
PASSWORD	
NOTES	

E

NAME/WEBSITE	
USERNAME/LOGIN	
PASSWORD	
NOTES	

NAME/WEBSITE	
USERNAME/LOGIN	
PASSWORD	
NOTES	

NAME/WEBSITE	
USERNAME/LOGIN	
PASSWORD	
NOTES	

NAME/WEBSITE	
USERNAME/LOGIN	
PASSWORD	
NOTES	

NAME/WEBSITE	
USERNAME/LOGIN	
PASSWORD	
NOTES	

NAME/WEBSITE	
USERNAME/LOGIN	
PASSWORD	
NOTES	

NAME/WEBSITE	
USERNAME/LOGIN	
PASSWORD	
NOTES	

NAME/WEBSITE	
USERNAME/LOGIN	
PASSWORD	
NOTES	

NAME/WEBSITE	
USERNAME/LOGIN	
PASSWORD	
NOTES	

NAME/WEBSITE	
USERNAME/LOGIN	
PASSWORD	
NOTES	

NAME/WEBSITE	
USERNAME/LOGIN	
PASSWORD	
NOTES	

NAME/WEBSITE	
USERNAME/LOGIN	
PASSWORD	
NOTES	

E

NAME/WEBSITE	
USERNAME/LOGIN	
PASSWORD	
NOTES	

NAME/WEBSITE	
USERNAME/LOGIN	
PASSWORD	
NOTES	

NAME/WEBSITE	
USERNAME/LOGIN	
PASSWORD	
NOTES	

NAME/WEBSITE	
USERNAME/LOGIN	
PASSWORD	
NOTES	

NAME/WEBSITE	
USERNAME/LOGIN	
PASSWORD	
NOTES	

NAME/WEBSITE	
USERNAME/LOGIN	
PASSWORD	
NOTES	

NAME/WEBSITE	
USERNAME/LOGIN	
PASSWORD	
NOTES	

NAME/WEBSITE	
USERNAME/LOGIN	
PASSWORD	
NOTES	

NAME/WEBSITE	
USERNAME/LOGIN	
PASSWORD	
NOTES	

NAME/WEBSITE	
USERNAME/LOGIN	
PASSWORD	
NOTES	

NAME/WEBSITE	
USERNAME/LOGIN	
PASSWORD	
NOTES	

NAME/WEBSITE	
USERNAME/LOGIN	
PASSWORD	
NOTES	

F
NAME/WEBSITE
USERNAME/LOGIN
PASSWORD
NOTES
NAME/WEBSITE
USERNAME/LOGIN
PASSWORD
NOTES
NAME/WEBSITE
USERNAME/LOGIN
PASSWORD
NOTES
NAME/WEBSITE
USERNAME/LOGIN
PASSWORD
NOTES
NAME/WEBSITE
USERNAME/LOGIN
PASSWORD
NOTES
NAME/WEBSITE
USERNAME/LOGIN
PASSWORD
NOTES

NAME/WEBSITE

USERNAME/LOGIN

PASSWORD

NOTES

NAME/WEBSITE

USERNAME/LOGIN

PASSWORD

NOTES

NAME/WEBSITE

USERNAME/LOGIN

PASSWORD

NOTES

NAME/WEBSITE

USERNAME/LOGIN

PASSWORD

NOTES

NAME/WEBSITE

USERNAME/LOGIN

PASSWORD

NOTES

NAME/WEBSITE

USERNAME/LOGIN

PASSWORD

NOTES

F
NAME/WEBSITE
USERNAME/LOGIN
PASSWORD
NOTES
NAME/WEBSITE
USERNAME/LOGIN
PASSWORD
NOTES
NAME/WEBSITE
USERNAME/LOGIN
PASSWORD
NOTES
NAME/WEBSITE
USERNAME/LOGIN
PASSWORD
NOTES
NAME/WEBSITE
USERNAME/LOGIN
PASSWORD
NOTES
NAME/WEBSITE
USERNAME/LOGIN
PASSWORD
NOTES

F	
NAME/WEBSITE	
USERNAME/LOGIN	
PASSWORD	
NOTES	
NAME/WEBSITE	
USERNAME/LOGIN	
PASSWORD	
NOTES	
NAME/WEBSITE	
USERNAME/LOGIN	
PASSWORD	
NOTES	
NAME/WEBSITE	
USERNAME/LOGIN	
PASSWORD	
NOTES	
NAME/WEBSITE	
USERNAME/LOGIN	
PASSWORD	
NOTES	
NAME/WEBSITE	
USERNAME/LOGIN	
PASSWORD	
NOTES	

G

NAME/WEBSITE	
USERNAME/LOGIN	
PASSWORD	
NOTES	

NAME/WEBSITE	
USERNAME/LOGIN	
PASSWORD	
NOTES	

NAME/WEBSITE	
USERNAME/LOGIN	
PASSWORD	
NOTES	

NAME/WEBSITE	
USERNAME/LOGIN	
PASSWORD	
NOTES	

NAME/WEBSITE	
USERNAME/LOGIN	
PASSWORD	
NOTES	

NAME/WEBSITE	
USERNAME/LOGIN	
PASSWORD	
NOTES	

NAME/WEBSITE

USERNAME/LOGIN

PASSWORD

NOTES

NAME/WEBSITE

USERNAME/LOGIN

PASSWORD

NOTES

NAME/WEBSITE

USERNAME/LOGIN

PASSWORD

NOTES

NAME/WEBSITE

USERNAME/LOGIN

PASSWORD

NOTES

NAME/WEBSITE

USERNAME/LOGIN

PASSWORD

NOTES

NAME/WEBSITE

USERNAME/LOGIN

PASSWORD

NOTES

G

NAME/WEBSITE

USERNAME/LOGIN

PASSWORD

NOTES

NAME/WEBSITE

USERNAME/LOGIN

PASSWORD

NOTES

NAME/WEBSITE

USERNAME/LOGIN

PASSWORD

NOTES

NAME/WEBSITE

USERNAME/LOGIN

PASSWORD

NOTES

NAME/WEBSITE

USERNAME/LOGIN

PASSWORD

NOTES

NAME/WEBSITE

USERNAME/LOGIN

PASSWORD

NOTES

NAME/WEBSITE	
USERNAME/LOGIN	
PASSWORD	
NOTES	

NAME/WEBSITE	
USERNAME/LOGIN	
PASSWORD	
NOTES	

NAME/WEBSITE	
USERNAME/LOGIN	
PASSWORD	
NOTES	

NAME/WEBSITE	
USERNAME/LOGIN	
PASSWORD	
NOTES	

NAME/WEBSITE	
USERNAME/LOGIN	
PASSWORD	
NOTES	

NAME/WEBSITE	
USERNAME/LOGIN	
PASSWORD	
NOTES	

H

NAME/WEBSITE	
USERNAME/LOGIN	
PASSWORD	
NOTES	

NAME/WEBSITE	
USERNAME/LOGIN	
PASSWORD	
NOTES	

NAME/WEBSITE	
USERNAME/LOGIN	
PASSWORD	
NOTES	

NAME/WEBSITE	
USERNAME/LOGIN	
PASSWORD	
NOTES	

NAME/WEBSITE	
USERNAME/LOGIN	
PASSWORD	
NOTES	

NAME/WEBSITE	
USERNAME/LOGIN	
PASSWORD	
NOTES	

NAME/WEBSITE

USERNAME/LOGIN

PASSWORD

NOTES

NAME/WEBSITE

USERNAME/LOGIN

PASSWORD

NOTES

NAME/WEBSITE

USERNAME/LOGIN

PASSWORD

NOTES

NAME/WEBSITE

USERNAME/LOGIN

PASSWORD

NOTES

NAME/WEBSITE

USERNAME/LOGIN

PASSWORD

NOTES

NAME/WEBSITE

USERNAME/LOGIN

PASSWORD

NOTES

H

NAME/WEBSITE	
USERNAME/LOGIN	
PASSWORD	
NOTES	
NAME/WEBSITE	
USERNAME/LOGIN	
PASSWORD	
NOTES	
NAME/WEBSITE	
USERNAME/LOGIN	
PASSWORD	
NOTES	
NAME/WEBSITE	
USERNAME/LOGIN	
PASSWORD	
NOTES	
NAME/WEBSITE	
USERNAME/LOGIN	
PASSWORD	
NOTES	
NAME/WEBSITE	
USERNAME/LOGIN	
PASSWORD	
NOTES	

H

NAME/WEBSITE

USERNAME/LOGIN

PASSWORD

NOTES

NAME/WEBSITE

USERNAME/LOGIN

PASSWORD

NOTES

NAME/WEBSITE

USERNAME/LOGIN

PASSWORD

NOTES

NAME/WEBSITE

USERNAME/LOGIN

PASSWORD

NOTES

NAME/WEBSITE

USERNAME/LOGIN

PASSWORD

NOTES

NAME/WEBSITE

USERNAME/LOGIN

PASSWORD

NOTES

I

NAME/WEBSITE

USERNAME/LOGIN

PASSWORD

NOTES

NAME/WEBSITE

USERNAME/LOGIN

PASSWORD

NOTES

NAME/WEBSITE

USERNAME/LOGIN

PASSWORD

NOTES

NAME/WEBSITE

USERNAME/LOGIN

PASSWORD

NOTES

NAME/WEBSITE

USERNAME/LOGIN

PASSWORD

NOTES

NAME/WEBSITE

USERNAME/LOGIN

PASSWORD

NOTES

NAME/WEBSITE	
USERNAME/LOGIN	
PASSWORD	
NOTES	
NAME/WEBSITE	
USERNAME/LOGIN	
PASSWORD	
NOTES	
NAME/WEBSITE	
USERNAME/LOGIN	
PASSWORD	
NOTES	
NAME/WEBSITE	
USERNAME/LOGIN	
PASSWORD	
NOTES	
NAME/WEBSITE	
USERNAME/LOGIN	
PASSWORD	
NOTES	
NAME/WEBSITE	
USERNAME/LOGIN	
PASSWORD	
NOTES	

	I
NAME/WEBSITE	
USERNAME/LOGIN	
PASSWORD	
NOTES	
NAME/WEBSITE	
USERNAME/LOGIN	
PASSWORD	
NOTES	
NAME/WEBSITE	
USERNAME/LOGIN	
PASSWORD	
NOTES	
NAME/WEBSITE	
USERNAME/LOGIN	
PASSWORD	
NOTES	
NAME/WEBSITE	
USERNAME/LOGIN	
PASSWORD	
NOTES	
NAME/WEBSITE	
USERNAME/LOGIN	
PASSWORD	
NOTES	

I

NAME/WEBSITE	
USERNAME/LOGIN	
PASSWORD	
NOTES	

NAME/WEBSITE	
USERNAME/LOGIN	
PASSWORD	
NOTES	

I

NAME/WEBSITE	
USERNAME/LOGIN	
PASSWORD	
NOTES	

NAME/WEBSITE	
USERNAME/LOGIN	
PASSWORD	
NOTES	

NAME/WEBSITE	
USERNAME/LOGIN	
PASSWORD	
NOTES	

NAME/WEBSITE	
USERNAME/LOGIN	
PASSWORD	
NOTES	

J

NAME/WEBSITE	
USERNAME/LOGIN	
PASSWORD	
NOTES	

NAME/WEBSITE	
USERNAME/LOGIN	
PASSWORD	
NOTES	

NAME/WEBSITE	
USERNAME/LOGIN	
PASSWORD	
NOTES	

NAME/WEBSITE	
USERNAME/LOGIN	
PASSWORD	
NOTES	

NAME/WEBSITE	
USERNAME/LOGIN	
PASSWORD	
NOTES	

NAME/WEBSITE	
USERNAME/LOGIN	
PASSWORD	
NOTES	

J

NAME/WEBSITE

USERNAME/LOGIN

PASSWORD

NOTES

NAME/WEBSITE

USERNAME/LOGIN

PASSWORD

NOTES

NAME/WEBSITE

USERNAME/LOGIN

PASSWORD

NOTES

NAME/WEBSITE

USERNAME/LOGIN

PASSWORD

NOTES

NAME/WEBSITE

USERNAME/LOGIN

PASSWORD

NOTES

NAME/WEBSITE

USERNAME/LOGIN

PASSWORD

NOTES

J
NAME/WEBSITE
USERNAME/LOGIN
PASSWORD
NOTES
NAME/WEBSITE
USERNAME/LOGIN
PASSWORD
NOTES
NAME/WEBSITE
USERNAME/LOGIN
PASSWORD
NOTES
NAME/WEBSITE
USERNAME/LOGIN
PASSWORD
NOTES
NAME/WEBSITE
USERNAME/LOGIN
PASSWORD
NOTES
NAME/WEBSITE
USERNAME/LOGIN
PASSWORD
NOTES

J

NAME/WEBSITE	
USERNAME/LOGIN	
PASSWORD	
NOTES	

NAME/WEBSITE	
USERNAME/LOGIN	
PASSWORD	
NOTES	

NAME/WEBSITE	
USERNAME/LOGIN	
PASSWORD	
NOTES	

NAME/WEBSITE	
USERNAME/LOGIN	
PASSWORD	
NOTES	

NAME/WEBSITE	
USERNAME/LOGIN	
PASSWORD	
NOTES	

NAME/WEBSITE	
USERNAME/LOGIN	
PASSWORD	
NOTES	

K

NAME/WEBSITE	
USERNAME/LOGIN	
PASSWORD	
NOTES	

NAME/WEBSITE	
USERNAME/LOGIN	
PASSWORD	
NOTES	

NAME/WEBSITE	
USERNAME/LOGIN	
PASSWORD	
NOTES	

NAME/WEBSITE	
USERNAME/LOGIN	
PASSWORD	
NOTES	

NAME/WEBSITE	
USERNAME/LOGIN	
PASSWORD	
NOTES	

NAME/WEBSITE	
USERNAME/LOGIN	
PASSWORD	
NOTES	

K

NAME/WEBSITE

USERNAME/LOGIN

PASSWORD

NOTES

NAME/WEBSITE

USERNAME/LOGIN

PASSWORD

NOTES

NAME/WEBSITE

USERNAME/LOGIN

PASSWORD

NOTES

NAME/WEBSITE

USERNAME/LOGIN

PASSWORD

NOTES

NAME/WEBSITE

USERNAME/LOGIN

PASSWORD

NOTES

NAME/WEBSITE

USERNAME/LOGIN

PASSWORD

NOTES

L

NAME/WEBSITE	
USERNAME/LOGIN	
PASSWORD	
NOTES	

NAME/WEBSITE	
USERNAME/LOGIN	
PASSWORD	
NOTES	

NAME/WEBSITE	
USERNAME/LOGIN	
PASSWORD	
NOTES	

NAME/WEBSITE	
USERNAME/LOGIN	
PASSWORD	
NOTES	

NAME/WEBSITE	
USERNAME/LOGIN	
PASSWORD	
NOTES	

NAME/WEBSITE	
USERNAME/LOGIN	
PASSWORD	
NOTES	

L

NAME/WEBSITE

USERNAME/LOGIN

PASSWORD

NOTES

NAME/WEBSITE

USERNAME/LOGIN

PASSWORD

NOTES

NAME/WEBSITE

USERNAME/LOGIN

PASSWORD

L

NOTES

NAME/WEBSITE

USERNAME/LOGIN

PASSWORD

NOTES

NAME/WEBSITE

USERNAME/LOGIN

PASSWORD

NOTES

NAME/WEBSITE

USERNAME/LOGIN

PASSWORD

NOTES

L	
NAME/WEBSITE	
USERNAME/LOGIN	
PASSWORD	
NOTES	
NAME/WEBSITE	
USERNAME/LOGIN	
PASSWORD	
NOTES	
NAME/WEBSITE	
USERNAME/LOGIN	
PASSWORD	
NOTES	
NAME/WEBSITE	
USERNAME/LOGIN	
PASSWORD	
NOTES	
NAME/WEBSITE	
USERNAME/LOGIN	
PASSWORD	
NOTES	
NAME/WEBSITE	
USERNAME/LOGIN	
PASSWORD	
NOTES	

NAME/WEBSITE

USERNAME/LOGIN

PASSWORD

NOTES

NAME/WEBSITE

USERNAME/LOGIN

PASSWORD

NOTES

NAME/WEBSITE

USERNAME/LOGIN

PASSWORD

NOTES

NAME/WEBSITE

USERNAME/LOGIN

PASSWORD

NOTES

NAME/WEBSITE

USERNAME/LOGIN

PASSWORD

NOTES

NAME/WEBSITE

USERNAME/LOGIN

PASSWORD

NOTES

M

NAME/WEBSITE	
USERNAME/LOGIN	
PASSWORD	
NOTES	

NAME/WEBSITE	
USERNAME/LOGIN	
PASSWORD	
NOTES	

NAME/WEBSITE	
USERNAME/LOGIN	
PASSWORD	
NOTES	

NAME/WEBSITE	
USERNAME/LOGIN	
PASSWORD	
NOTES	

NAME/WEBSITE	
USERNAME/LOGIN	
PASSWORD	
NOTES	

NAME/WEBSITE	
USERNAME/LOGIN	
PASSWORD	
NOTES	

NAME/WEBSITE	
USERNAME/LOGIN	
PASSWORD	
NOTES	

NAME/WEBSITE	
USERNAME/LOGIN	
PASSWORD	
NOTES	

NAME/WEBSITE	
USERNAME/LOGIN	
PASSWORD	
NOTES	

NAME/WEBSITE	
USERNAME/LOGIN	
PASSWORD	
NOTES	

NAME/WEBSITE	
USERNAME/LOGIN	
PASSWORD	
NOTES	

NAME/WEBSITE	
USERNAME/LOGIN	
PASSWORD	
NOTES	

M

NAME/WEBSITE	
USERNAME/LOGIN	
PASSWORD	
NOTES	

NAME/WEBSITE	
USERNAME/LOGIN	
PASSWORD	
NOTES	

NAME/WEBSITE	
USERNAME/LOGIN	
PASSWORD	
NOTES	

NAME/WEBSITE	
USERNAME/LOGIN	
PASSWORD	
NOTES	

NAME/WEBSITE	
USERNAME/LOGIN	
PASSWORD	
NOTES	

NAME/WEBSITE	
USERNAME/LOGIN	
PASSWORD	
NOTES	

M	
NAME/WEBSITE	
USERNAME/LOGIN	
PASSWORD	
NOTES	
NAME/WEBSITE	
USERNAME/LOGIN	
PASSWORD	
NOTES	
NAME/WEBSITE	
USERNAME/LOGIN	
PASSWORD	
NOTES	
NAME/WEBSITE	
USERNAME/LOGIN	
PASSWORD	
NOTES	
NAME/WEBSITE	
USERNAME/LOGIN	
PASSWORD	
NOTES	
NAME/WEBSITE	
USERNAME/LOGIN	
PASSWORD	
NOTES	

N

NAME/WEBSITE	
USERNAME/LOGIN	
PASSWORD	
NOTES	

NAME/WEBSITE	
USERNAME/LOGIN	
PASSWORD	
NOTES	

NAME/WEBSITE	
USERNAME/LOGIN	
PASSWORD	
NOTES	

NAME/WEBSITE	
USERNAME/LOGIN	
PASSWORD	
NOTES	

NAME/WEBSITE	
USERNAME/LOGIN	
PASSWORD	
NOTES	

NAME/WEBSITE	
USERNAME/LOGIN	
PASSWORD	
NOTES	

NAME/WEBSITE	
USERNAME/LOGIN	
PASSWORD	
NOTES	

NAME/WEBSITE	
USERNAME/LOGIN	
PASSWORD	
NOTES	

NAME/WEBSITE	
USERNAME/LOGIN	
PASSWORD	
NOTES	

NAME/WEBSITE	
USERNAME/LOGIN	
PASSWORD	
NOTES	

NAME/WEBSITE	
USERNAME/LOGIN	
PASSWORD	
NOTES	

NAME/WEBSITE	
USERNAME/LOGIN	
PASSWORD	
NOTES	

N

NAME/WEBSITE	
USERNAME/LOGIN	
PASSWORD	
NOTES	

NAME/WEBSITE	
USERNAME/LOGIN	
PASSWORD	
NOTES	

NAME/WEBSITE	
USERNAME/LOGIN	
PASSWORD	
NOTES	

NAME/WEBSITE	
USERNAME/LOGIN	
PASSWORD	
NOTES	

NAME/WEBSITE	
USERNAME/LOGIN	
PASSWORD	
NOTES	

NAME/WEBSITE	
USERNAME/LOGIN	
PASSWORD	
NOTES	

NAME/WEBSITE	
USERNAME/LOGIN	
PASSWORD	
NOTES	

NAME/WEBSITE	
USERNAME/LOGIN	
PASSWORD	
NOTES	

NAME/WEBSITE	
USERNAME/LOGIN	
PASSWORD	
NOTES	

NAME/WEBSITE	
USERNAME/LOGIN	
PASSWORD	
NOTES	

NAME/WEBSITE	
USERNAME/LOGIN	
PASSWORD	
NOTES	

NAME/WEBSITE	
USERNAME/LOGIN	
PASSWORD	
NOTES	

	O
NAME/WEBSITE	
USERNAME/LOGIN	
PASSWORD	
NOTES	
NAME/WEBSITE	
USERNAME/LOGIN	
PASSWORD	
NOTES	
NAME/WEBSITE	
USERNAME/LOGIN	
PASSWORD	
NOTES	
NAME/WEBSITE	
USERNAME/LOGIN	
PASSWORD	
NOTES	
NAME/WEBSITE	
USERNAME/LOGIN	
PASSWORD	
NOTES	
NAME/WEBSITE	
USERNAME/LOGIN	
PASSWORD	
NOTES	

NAME/WEBSITE

USERNAME/LOGIN

PASSWORD

NOTES

NAME/WEBSITE

USERNAME/LOGIN

PASSWORD

NOTES

NAME/WEBSITE

USERNAME/LOGIN

PASSWORD

NOTES

NAME/WEBSITE

USERNAME/LOGIN

PASSWORD

NOTES

NAME/WEBSITE

USERNAME/LOGIN

PASSWORD

NOTES

NAME/WEBSITE

USERNAME/LOGIN

PASSWORD

NOTES

NAME/WEBSITE	
USERNAME/LOGIN	
PASSWORD	
NOTES	
NAME/WEBSITE	
USERNAME/LOGIN	
PASSWORD	
NOTES	
NAME/WEBSITE	
USERNAME/LOGIN	
PASSWORD	
NOTES	
NAME/WEBSITE	
USERNAME/LOGIN	
PASSWORD	
NOTES	
NAME/WEBSITE	
USERNAME/LOGIN	
PASSWORD	
NOTES	
NAME/WEBSITE	
USERNAME/LOGIN	
PASSWORD	
NOTES	

NAME/WEBSITE

USERNAME/LOGIN

PASSWORD

NOTES

NAME/WEBSITE

USERNAME/LOGIN

PASSWORD

NOTES

NAME/WEBSITE

USERNAME/LOGIN

PASSWORD

NOTES

NAME/WEBSITE

USERNAME/LOGIN

PASSWORD

NOTES

NAME/WEBSITE

USERNAME/LOGIN

PASSWORD

NOTES

NAME/WEBSITE

USERNAME/LOGIN

PASSWORD

NOTES

P

NAME/WEBSITE	
USERNAME/LOGIN	
PASSWORD	
NOTES	

NAME/WEBSITE	
USERNAME/LOGIN	
PASSWORD	
NOTES	

NAME/WEBSITE	
USERNAME/LOGIN	
PASSWORD	
NOTES	

NAME/WEBSITE	
USERNAME/LOGIN	
PASSWORD	
NOTES	

NAME/WEBSITE	
USERNAME/LOGIN	
PASSWORD	
NOTES	

NAME/WEBSITE	
USERNAME/LOGIN	
PASSWORD	
NOTES	

NAME/WEBSITE	
USERNAME/LOGIN	
PASSWORD	
NOTES	

NAME/WEBSITE	
USERNAME/LOGIN	
PASSWORD	
NOTES	

NAME/WEBSITE	
USERNAME/LOGIN	
PASSWORD	
NOTES	

NAME/WEBSITE	
USERNAME/LOGIN	
PASSWORD	
NOTES	

NAME/WEBSITE	
USERNAME/LOGIN	
PASSWORD	
NOTES	

NAME/WEBSITE	
USERNAME/LOGIN	
PASSWORD	
NOTES	

P

NAME/WEBSITE	
USERNAME/LOGIN	
PASSWORD	
NOTES	

NAME/WEBSITE	
USERNAME/LOGIN	
PASSWORD	
NOTES	

NAME/WEBSITE	
USERNAME/LOGIN	
PASSWORD	
NOTES	

NAME/WEBSITE	
USERNAME/LOGIN	
PASSWORD	
NOTES	

NAME/WEBSITE	
USERNAME/LOGIN	
PASSWORD	
NOTES	

NAME/WEBSITE	
USERNAME/LOGIN	
PASSWORD	
NOTES	

NAME/WEBSITE

USERNAME/LOGIN

PASSWORD

NOTES

NAME/WEBSITE

USERNAME/LOGIN

PASSWORD

NOTES

NAME/WEBSITE

USERNAME/LOGIN

PASSWORD

NOTES

NAME/WEBSITE

USERNAME/LOGIN

PASSWORD

NOTES

NAME/WEBSITE

USERNAME/LOGIN

PASSWORD

NOTES

NAME/WEBSITE

USERNAME/LOGIN

PASSWORD

NOTES

Q

NAME/WEBSITE	
USERNAME/LOGIN	
PASSWORD	
NOTES	

NAME/WEBSITE	
USERNAME/LOGIN	
PASSWORD	
NOTES	

NAME/WEBSITE	
USERNAME/LOGIN	
PASSWORD	
NOTES	

NAME/WEBSITE	
USERNAME/LOGIN	
PASSWORD	
NOTES	

NAME/WEBSITE	
USERNAME/LOGIN	
PASSWORD	
NOTES	

NAME/WEBSITE	
USERNAME/LOGIN	
PASSWORD	
NOTES	

NAME/WEBSITE	
USERNAME/LOGIN	
PASSWORD	
NOTES	

NAME/WEBSITE	
USERNAME/LOGIN	
PASSWORD	
NOTES	

NAME/WEBSITE	
USERNAME/LOGIN	
PASSWORD	
NOTES	

NAME/WEBSITE	
USERNAME/LOGIN	
PASSWORD	
NOTES	

NAME/WEBSITE	
USERNAME/LOGIN	
PASSWORD	
NOTES	

NAME/WEBSITE	
USERNAME/LOGIN	
PASSWORD	
NOTES	

R

NAME/WEBSITE	
USERNAME/LOGIN	
PASSWORD	
NOTES	

NAME/WEBSITE	
USERNAME/LOGIN	
PASSWORD	
NOTES	

NAME/WEBSITE	
USERNAME/LOGIN	
PASSWORD	
NOTES	

NAME/WEBSITE	
USERNAME/LOGIN	
PASSWORD	
NOTES	

NAME/WEBSITE	
USERNAME/LOGIN	
PASSWORD	
NOTES	

NAME/WEBSITE	
USERNAME/LOGIN	
PASSWORD	
NOTES	

R

NAME/WEBSITE

USERNAME/LOGIN

PASSWORD

NOTES

NAME/WEBSITE

USERNAME/LOGIN

PASSWORD

NOTES

NAME/WEBSITE

USERNAME/LOGIN

PASSWORD

NOTES

NAME/WEBSITE

USERNAME/LOGIN

PASSWORD

NOTES

NAME/WEBSITE

USERNAME/LOGIN

PASSWORD

NOTES

NAME/WEBSITE

USERNAME/LOGIN

PASSWORD

NOTES

R

NAME/WEBSITE	
USERNAME/LOGIN	
PASSWORD	
NOTES	

NAME/WEBSITE	
USERNAME/LOGIN	
PASSWORD	
NOTES	

NAME/WEBSITE	
USERNAME/LOGIN	
PASSWORD	
NOTES	

NAME/WEBSITE	
USERNAME/LOGIN	
PASSWORD	
NOTES	

NAME/WEBSITE	
USERNAME/LOGIN	
PASSWORD	
NOTES	

NAME/WEBSITE	
USERNAME/LOGIN	
PASSWORD	
NOTES	

R
NAME/WEBSITE
USERNAME/LOGIN
PASSWORD
NOTES
NAME/WEBSITE
USERNAME/LOGIN
PASSWORD
NOTES
NAME/WEBSITE
USERNAME/LOGIN
PASSWORD
NOTES
NAME/WEBSITE
USERNAME/LOGIN
PASSWORD
NOTES
NAME/WEBSITE
USERNAME/LOGIN
PASSWORD
NOTES
NAME/WEBSITE
USERNAME/LOGIN
PASSWORD
NOTES

S	
NAME/WEBSITE	
USERNAME/LOGIN	
PASSWORD	
NOTES	
NAME/WEBSITE	
USERNAME/LOGIN	
PASSWORD	
NOTES	
NAME/WEBSITE	
USERNAME/LOGIN	
PASSWORD	
NOTES	
NAME/WEBSITE	
USERNAME/LOGIN	
PASSWORD	
NOTES	
NAME/WEBSITE	
USERNAME/LOGIN	
PASSWORD	
NOTES	
NAME/WEBSITE	
USERNAME/LOGIN	
PASSWORD	
NOTES	

S	
NAME/WEBSITE	
USERNAME/LOGIN	
PASSWORD	
NOTES	
NAME/WEBSITE	
USERNAME/LOGIN	
PASSWORD	
NOTES	
NAME/WEBSITE	
USERNAME/LOGIN	
PASSWORD	
NOTES	
NAME/WEBSITE	
USERNAME/LOGIN	
PASSWORD	
NOTES	
NAME/WEBSITE	
USERNAME/LOGIN	
PASSWORD	
NOTES	
NAME/WEBSITE	
USERNAME/LOGIN	
PASSWORD	
NOTES	

S	
NAME/WEBSITE	
USERNAME/LOGIN	
PASSWORD	
NOTES	
NAME/WEBSITE	
USERNAME/LOGIN	
PASSWORD	
NOTES	
NAME/WEBSITE	
USERNAME/LOGIN	
PASSWORD	
NOTES	
NAME/WEBSITE	
USERNAME/LOGIN	
PASSWORD	
NOTES	
NAME/WEBSITE	
USERNAME/LOGIN	
PASSWORD	
NOTES	
NAME/WEBSITE	
USERNAME/LOGIN	
PASSWORD	
NOTES	

NAME/WEBSITE	
USERNAME/LOGIN	
PASSWORD	
NOTES	

NAME/WEBSITE	
USERNAME/LOGIN	
PASSWORD	
NOTES	

NAME/WEBSITE	
USERNAME/LOGIN	
PASSWORD	
NOTES	

NAME/WEBSITE	
USERNAME/LOGIN	
PASSWORD	
NOTES	

NAME/WEBSITE	
USERNAME/LOGIN	
PASSWORD	
NOTES	

NAME/WEBSITE	
USERNAME/LOGIN	
PASSWORD	
NOTES	

T

NAME/WEBSITE	
USERNAME/LOGIN	
PASSWORD	
NOTES	

NAME/WEBSITE	
USERNAME/LOGIN	
PASSWORD	
NOTES	

NAME/WEBSITE	
USERNAME/LOGIN	
PASSWORD	
NOTES	

NAME/WEBSITE	
USERNAME/LOGIN	
PASSWORD	
NOTES	

NAME/WEBSITE	
USERNAME/LOGIN	
PASSWORD	
NOTES	

NAME/WEBSITE	
USERNAME/LOGIN	
PASSWORD	
NOTES	

T

NAME/WEBSITE	
USERNAME/LOGIN	
PASSWORD	
NOTES	

NAME/WEBSITE	
USERNAME/LOGIN	
PASSWORD	
NOTES	

NAME/WEBSITE	
USERNAME/LOGIN	
PASSWORD	
NOTES	

NAME/WEBSITE	
USERNAME/LOGIN	
PASSWORD	
NOTES	

NAME/WEBSITE	
USERNAME/LOGIN	
PASSWORD	
NOTES	

NAME/WEBSITE	
USERNAME/LOGIN	
PASSWORD	
NOTES	

T

NAME/WEBSITE

USERNAME/LOGIN

PASSWORD

NOTES

NAME/WEBSITE

USERNAME/LOGIN

PASSWORD

NOTES

NAME/WEBSITE

USERNAME/LOGIN

PASSWORD

NOTES

NAME/WEBSITE

USERNAME/LOGIN

PASSWORD

NOTES

NAME/WEBSITE

USERNAME/LOGIN

PASSWORD

NOTES

NAME/WEBSITE

USERNAME/LOGIN

PASSWORD

NOTES

T

NAME/WEBSITE

USERNAME/LOGIN

PASSWORD

NOTES

NAME/WEBSITE

USERNAME/LOGIN

PASSWORD

NOTES

NAME/WEBSITE

USERNAME/LOGIN

PASSWORD

NOTES

NAME/WEBSITE

USERNAME/LOGIN

PASSWORD

NOTES

NAME/WEBSITE

USERNAME/LOGIN

PASSWORD

NOTES

NAME/WEBSITE

USERNAME/LOGIN

PASSWORD

NOTES

T

U

NAME/WEBSITE	
USERNAME/LOGIN	
PASSWORD	
NOTES	

NAME/WEBSITE	
USERNAME/LOGIN	
PASSWORD	
NOTES	

NAME/WEBSITE	
USERNAME/LOGIN	
PASSWORD	
NOTES	

NAME/WEBSITE	
USERNAME/LOGIN	
PASSWORD	
NOTES	

NAME/WEBSITE	
USERNAME/LOGIN	
PASSWORD	
NOTES	

NAME/WEBSITE	
USERNAME/LOGIN	
PASSWORD	
NOTES	

U

NAME/WEBSITE	
USERNAME/LOGIN	
PASSWORD	
NOTES	

NAME/WEBSITE	
USERNAME/LOGIN	
PASSWORD	
NOTES	

NAME/WEBSITE	
USERNAME/LOGIN	
PASSWORD	
NOTES	

NAME/WEBSITE	
USERNAME/LOGIN	
PASSWORD	
NOTES	

NAME/WEBSITE	
USERNAME/LOGIN	
PASSWORD	
NOTES	

NAME/WEBSITE	
USERNAME/LOGIN	
PASSWORD	
NOTES	

V

NAME/WEBSITE	
USERNAME/LOGIN	
PASSWORD	
NOTES	

NAME/WEBSITE	
USERNAME/LOGIN	
PASSWORD	
NOTES	

NAME/WEBSITE	
USERNAME/LOGIN	
PASSWORD	
NOTES	

NAME/WEBSITE	
USERNAME/LOGIN	
PASSWORD	
NOTES	

NAME/WEBSITE	
USERNAME/LOGIN	
PASSWORD	
NOTES	

NAME/WEBSITE	
USERNAME/LOGIN	
PASSWORD	
NOTES	

V

NAME/WEBSITE	
USERNAME/LOGIN	
PASSWORD	
NOTES	
NAME/WEBSITE	
USERNAME/LOGIN	
PASSWORD	
NOTES	
NAME/WEBSITE	
USERNAME/LOGIN	
PASSWORD	
NOTES	
NAME/WEBSITE	
USERNAME/LOGIN	
PASSWORD	
NOTES	
NAME/WEBSITE	
USERNAME/LOGIN	
PASSWORD	
NOTES	
NAME/WEBSITE	
USERNAME/LOGIN	
PASSWORD	
NOTES	

V

W

NAME/WEBSITE	
USERNAME/LOGIN	
PASSWORD	
NOTES	

NAME/WEBSITE	
USERNAME/LOGIN	
PASSWORD	
NOTES	

NAME/WEBSITE	
USERNAME/LOGIN	
PASSWORD	
NOTES	

NAME/WEBSITE	
USERNAME/LOGIN	
PASSWORD	
NOTES	

NAME/WEBSITE	
USERNAME/LOGIN	
PASSWORD	
NOTES	

NAME/WEBSITE	
USERNAME/LOGIN	
PASSWORD	
NOTES	

NAME/WEBSITE

USERNAME/LOGIN

PASSWORD

NOTES

NAME/WEBSITE

USERNAME/LOGIN

PASSWORD

NOTES

NAME/WEBSITE

USERNAME/LOGIN

PASSWORD

NOTES

NAME/WEBSITE

USERNAME/LOGIN

PASSWORD

NOTES

NAME/WEBSITE

USERNAME/LOGIN

PASSWORD

NOTES

NAME/WEBSITE

USERNAME/LOGIN

PASSWORD

NOTES

X	
NAME/WEBSITE	
USERNAME/LOGIN	
PASSWORD	
NOTES	
NAME/WEBSITE	
USERNAME/LOGIN	
PASSWORD	
NOTES	
NAME/WEBSITE	
USERNAME/LOGIN	
PASSWORD	
NOTES	
NAME/WEBSITE	
USERNAME/LOGIN	
PASSWORD	
NOTES	
NAME/WEBSITE	
USERNAME/LOGIN	
PASSWORD	
NOTES	
NAME/WEBSITE	
USERNAME/LOGIN	
PASSWORD	
NOTES	

X	
NAME/WEBSITE	
USERNAME/LOGIN	
PASSWORD	
NOTES	
NAME/WEBSITE	
USERNAME/LOGIN	
PASSWORD	
NOTES	
NAME/WEBSITE	
USERNAME/LOGIN	
PASSWORD	
NOTES	
NAME/WEBSITE	
USERNAME/LOGIN	
PASSWORD	
NOTES	
NAME/WEBSITE	
USERNAME/LOGIN	
PASSWORD	
NOTES	
NAME/WEBSITE	
USERNAME/LOGIN	
PASSWORD	
NOTES	

X

Y

NAME/WEBSITE	
USERNAME/LOGIN	
PASSWORD	
NOTES	

NAME/WEBSITE	
USERNAME/LOGIN	
PASSWORD	
NOTES	

NAME/WEBSITE	
USERNAME/LOGIN	
PASSWORD	
NOTES	

NAME/WEBSITE	
USERNAME/LOGIN	
PASSWORD	
NOTES	

NAME/WEBSITE	
USERNAME/LOGIN	
PASSWORD	
NOTES	

NAME/WEBSITE	
USERNAME/LOGIN	
PASSWORD	
NOTES	

NAME/WEBSITE	
USERNAME/LOGIN	
PASSWORD	
NOTES	

NAME/WEBSITE	
USERNAME/LOGIN	
PASSWORD	
NOTES	

NAME/WEBSITE	
USERNAME/LOGIN	
PASSWORD	
NOTES	

NAME/WEBSITE	
USERNAME/LOGIN	
PASSWORD	
NOTES	

NAME/WEBSITE	
USERNAME/LOGIN	
PASSWORD	
NOTES	

NAME/WEBSITE	
USERNAME/LOGIN	
PASSWORD	
NOTES	

Z

NAME/WEBSITE	
USERNAME/LOGIN	
PASSWORD	
NOTES	
NAME/WEBSITE	
USERNAME/LOGIN	
PASSWORD	
NOTES	
NAME/WEBSITE	
USERNAME/LOGIN	
PASSWORD	
NOTES	
NAME/WEBSITE	
USERNAME/LOGIN	
PASSWORD	
NOTES	
NAME/WEBSITE	
USERNAME/LOGIN	
PASSWORD	
NOTES	
NAME/WEBSITE	
USERNAME/LOGIN	
PASSWORD	
NOTES	

NAME/WEBSITE

USERNAME/LOGIN

PASSWORD

NOTES

NAME/WEBSITE

USERNAME/LOGIN

PASSWORD

NOTES

NAME/WEBSITE

USERNAME/LOGIN

PASSWORD

NOTES

NAME/WEBSITE

USERNAME/LOGIN

PASSWORD

NOTES

NAME/WEBSITE

USERNAME/LOGIN

PASSWORD

NOTES

NAME/WEBSITE

USERNAME/LOGIN

PASSWORD

NOTES

NAME/WEBSITE

USERNAME/LOGIN

PASSWORD

NOTES

NOTES